On Telepathy

by Colin Griffith

I0486573

ISBN 978-1-387-60604-7

Introduction:

At an undisclosed location

"Tell me the first sequence of numbers you can think of," asked the man on the telephone. "4608" was the response. Those were the numbers transmitted via thought, across America, and across the Atlantic to an anonymous receiver in Europe. This young European repeated those exact numbers that I had written down to the telephone's operator, who was in the room with myself, as I attempted to send messages via thought alone to persons I knew around the world.

It was time to try another sequence with another person, this time with a friend up the coast in Washington. "2167" came the answer. It was exactly what I had written down.

We moved back to the first caller. The telephone operator left the room and then came back to ask me, "What are the first words that come to your mind?"

"Love, marriage, then comes the baby carriage."

Then, he showed me the paper where he had written; "Love, marriage, baby carriage." It was the words the caller in Europe had given him over the telephone.

Test complete.

This test relies on honesty. This entire scientific enterprise would be completely undermined should it become known that the participants had any prior communication regarding sequences of words or numbers. Nor does it conclusively prove telepathy even assuming complete honesty. This is because the participants did know each other prior to the experiment and thus it could be said that they simply knew each other so well that they could guess the correct sequence. However, this solution seems improbable compared to telepathic communication. Thus, there is sufficient evidence to pursue further examination of the subject of telepathy. This must include a working theoretical framework for telepathy in order to be taken seriously. Finding such a solution must surely lay in the mysteries of quantum mechanics.

Analysis and Abstract:

Many people speak of hearing voices in their head. To be clear they are not hearing voices, as in having hallucinations, but rather are witnessing an internal dialogue where the words of others appear directly within their frame of reference. These voices can be of persons present or absent, near or far. If it could be shown that the words of a person in another's mind were reflective of that person's actual thoughts or inner state in anyway, such communication would have to be considered telepathic.

There must be a theoretical basis for telepathy that does not break the rules of physics expressed in General Relativity and the Standard Model. There cannot be faster than light travel. However, information is certainly being translated as particles, most likely massless and in that case traveling at the speed of light. The question remains that if the distances were large enough to make a difference, would this communication be possible instantaneously. This would certainly indicate quantum tunneling and very likely quantum entanglement. For now, our experiment leaves the possibility that information is traveling at or very close to the speed of light, which seems to be the simplest explanation. The key is to look for symmetry between two or more persons streams of thought and frames of reference. If we could see invariance between two individuals thought patterns, we would be looking at a theoretical explanation for short range telepathy.

Thoughts are not brain matter nor are they chemicals initiating signals. Thoughts are the culmination of the the electromagnetic fields produced by the traveling charges within the brain. Brain states are thought patterns and they exist as waves with amplitudes and frequencies. The patterns within the waves lead to streams of thought. If these waves were to find a path to another field (aka another brain) there could be communication at the speed of light. Hence, classical field theory and Maxwell's equations lie at the heart of this project. The mathematics will almost certainly involve path integrals in order to achieve an understanding of the quantum electro dynamics.

There is one great question at the heart of our discussion of telepathy. The question is one of who is the active telepath behind thought messages? Is it the person sending the thought message or the individual reading the thoughts? It would appear to me that a typical telepath is sensitive to the thought patterns of other people. Within that group there appear to be individuals who are capable of sending a thought message. My assumption is that only persons who are also telepathic are capable of accurately deciphering those messages. To most persons, the message would likely appear as a stream of wordless consciousness and/or waves of perception. Thus, the thought message itself represents a bilateral experience of information along with the willful intent to communicate at both ends of this process, sending and receiving the message.

Addendum: Dream Sharing

The subject of dream content is critical to our exploration of telepathy. Specifically, we will want to take a look at the phenomena of dream sharing. Dream sharing occurs when two or more dreamers remember similar dream content upon waking. They may even remember each other's presence(s), conversations within the dream, and the major dream events. This is strong evidence of telepathic communication. The question remains, are the essential pieces of evidence instantaneously occurring in multiple frames of reference at the same time? Or are they communicated by massless particles at the speed of light? If this were the case, which I believe it is, then the events could not be remembered simultaneously. The distance between dreamers, however small, would not be negligible. Perhaps it is for this reason that dream sharing normally occurs with people sleeping near each other. However, I have experienced the sharing of dream content with people thousands of miles away at the time. Thus, the theory of special relativity would be crucial to understanding the phenomena. When examined closely, one realizes that they would receive the input of events within the dream at a time relative to their perceptions of the other dreamers. Hence, relatively speaking they would perceive simultaneity. Yet, I am quite sure that if we could time each dreamers experience of critical events from an outside perspective, we would not measure simultaneity.

Addendum: Remote Viewing

It should be noted that after the completion of the sequence testing a remote viewing experiment was performed with the same participants. The participants were unable in any of three trials to describe the setting or clothes of the other callers. Seeing as there was such success with telepathic sequence communication the complete failure of remote viewing with the same participants serves as strong evidence that remote viewing is no more telepathically effective than guess work.

I theorize that remote viewing does not work, or at the very least is significantly more difficult, is because exchanging information between two sets of mirror neurons allows you access to thought patterns, and not senses. It certainly would not give one the ability to view another person from an external perspective as remote viewers claim to be able to do. For this reason, remote viewing is not explored further, however its discussion so far is necessary for the overall project.

Addendum: String Theory

When we are discussing massless particles and wave equations translating information, what we are ultimately discussing is energy. What we know about energy is that it exists in discrete packets called quanta. Most forms of string theory decode the universe by imagining vibrating strings creating quarks which in tern constitute subatomic particles. At this level, raw information would appear to consist of pure energy. The key with any string theory is its symmetry. If strings were found to be symmetrical across spacetime then we could imagine electromagnetic waves propelling information at light speed across the galaxy. We would need to see that there is invariance in the wave equations of two or more minds. With work, we may be able to offer string theory as a plausible explanation for dream sharing and other telepathic communication.

Observations:

Dream Sharing

Dream sharing occurs when multiple observers report having the same dream, and even of seeing one another within the dream along with recall of similar events. This requires the transmission of information. That information is being communicated without any visual or auditory cues. This would seem to require the interaction of the observers mental states. The only way that such interaction would be possible would be at the quantum level. We now look at equations from standard quantum mechanics to see if they can be an effective model for dream sharing.

To begin our theoretical treatment of dream sharing we shall examine some equations from Susskind's The Theoretical Minimum: Quantum Mechanics. On page 164 we see that,

$$\alpha_u | u\} + \alpha_d | d\}$$
$$\beta_u | u\rangle + \beta_d | d\rangle$$

$$| Product \; State\rangle = \{\alpha_u | u\} + \alpha_d | d\}\} \otimes \{\beta_u | u\rangle + B_d | d\rangle\}$$

$$| Product \; State\rangle = \alpha_u \beta_u | uu\rangle + \alpha_u \beta_d | ud\rangle + \alpha_d \beta_u | du\rangle + \alpha_d \beta_d | dd\rangle$$

The first two relationships provide the spin of two separate particles alpha and beta. Each of these particles from separate states has an effect on the other. Thus, they produce a combined state. We calculate this by taking the tensor product of the two particles and calling it the Product State. The Product State is the result of all the components of the two particles with different spin and gives us information about the resulting combine system. Thus, there is a precedent for combining the states of two particles already within quantum mechanics. The Product State can be any real value but 0. If the Product State of two particles exists then always exists, as it must there should also be a Product State resulting from all the quantum particles making up the minds of multiple observers. The interactions would have to be close, but if all the quantum particles of one mind were treated as one particle, we could model the situation using the above equations and assume that the product state would exist even at great distances. This could account for dream sharing between observers not in the same room, or even the same town.

Perhaps more importantly, dream sharing does not allow for determinism and neither does quantum mechanics. Dreams are pure thought, and give evidence for a free reigning will bestowed upon each person. Dream sharing is the uniting of those wills. All of those wills involved must be free, for there is no other way that the different minds could compute and comprehend the actions of one will within the dream, much less all of them together. This is shown in quantum mechanics as uncertainty. The General Uncertainty Principle is derived in Susskind's book on Quantum Mechanics on page 146. I present this to you below.

$$2|X||Y| \geq |\langle X| \ Y\rangle + \langle Y| \ X\rangle|$$

$$|X\rangle = A| \ \Psi\rangle$$
$$|Y\rangle = iB| \ \Psi\rangle$$

$$2\sqrt{\langle A^2\rangle \langle B^2\rangle} \geq |\langle \Psi|AB| \ \Psi\rangle - \langle \Psi|BA| \ \Psi\rangle|$$

$$2\sqrt{\langle A^2\rangle \langle B^2\rangle} \geq |\langle \Psi|[A,B]| \ \Psi\rangle|$$

The first equation is the Cauchy-Schwarz inequality. It relates the two vectors X and Y. In other words, it connects the changes in position in two dimensions for a particle. X and Y are then defined as wave functions using the character Psi. These definitions are then substituted into the original inequality to produce the second inequality. The final inequality is then derived from that. This inequality is known as the General Uncertainty Principle. It states that the product of the uncertainties for A and B "cannot be smaller than half the magnitude of the expectation value of the commutator." If the product of the uncertainties can never be zero, it is impossible to know the particle's position and momentum exactly. In other words, there is always uncertainty.

If A and B have expectation values of O, the following two equations can be applied. Then, the final inequality above can be rewritten as the inequality below. This is the General Uncertainty Principle in mathematical form.

$$\langle A^2 \rangle = (\Delta A)^2$$
$$\langle B^2 \rangle = (\Delta B)^2$$

$$\Delta A \Delta B \geq \frac{1}{2} |\langle \Psi | [A,B] | \Psi \rangle|$$

Suppose a linear operator M acting on the space states of the composite system. The matrix elements of M are expressed as below. From page 161

$$|\Psi\rangle = \sum_{a,b} \psi(a,b)|\,ab\rangle$$

$$\langle a'b'|M|\,ab\rangle = M_{a'b',ab}$$

$$|\Psi\rangle = \sum_{a,b} \psi(a,b)|\,ab\rangle$$

On page 161, we are told that, "Now that we have the basis vectors, any linear superposition of them is allowed. Thus, any state in the compounded state can be expanded as" the final equation. This equation tells us that the wave function is determined by the sum of all the velocities for the two particles a and b. This is shown above.

Perhaps, the biggest question with dream sharing is; are the quantum fields entangled?

On page 166, Susskind gives us an example of maximally entangled state known as the "singlet state". The equation for a singlet state is shown below.

$$| \ sing \rangle = \frac{1}{\sqrt{2}} (| \ ud \rangle - | \ du \rangle)$$

The singlet state cannot be written as a product state. The singlet gives us everything about the combined system of two spins, but nothing about the individual spins themselves. This relates to the possibility of entangled dreamers by showing how the two fields produce one combined state. In our case, one state would mean one dream, being viewed by multiple people. This truly is what is meant by dream sharing.

Telepathic Communication:

Telepathic communication involves the transmission of information at a distance. In our experiment the subjects were separated by hundreds of miles. All that is for the test necessary is that the subjects are in different rooms so that body communication and other visual cues can be ruled out. That being said, it seems implausible that the the thought states of the subjects, their electromagnetic fields, are interacting without the interference of every other persons' mental fields. Such a situation would result in sheer chaos. Thus, the information must be transferred by particles, in which case the maximum speed even for a massless particle is C, the speed of light. We can then imagine that if we were to test telepathy across the distances of outer space, their would be a time delay. If the time it took for messages to travel was exactly the speed of light, we would be dealing with a massless particle. It would also seem, that if it were a tiny, but still massive particle, it would have been discovered already and its interactions understood. A particle with mass, such as an electron is tied to its field and cannot jump from mind to mind across great distances without affecting the core structure - the atom - that is a part of. Thus, a massless particle is the best candidate for telepathic communication and therefore we can hypothesize that thoughts exist in the physical world as massless particles.

In order to analyze such interactions we need to make use of what is known as a "Path Integral." Path Integrals are a sum over all available paths. We do not know which path a particle might take we only know the beginning and end points. For this reason, we need to calculate an integral of all potential paths. This process is best explained in Richard Feynman's book, <u>Quantum Mechanics and Path Integrals</u>.

We begin with a sum without integration. In easier examples, we can use the substitution below from page 34 of Feynman's book to calculate a sum of all paths.

$$\ddot{x} = \frac{1}{\varepsilon^2}(x_i + 1 - 2x_i + x_{i-1})$$

In more difficult cases integration is necessary. Thus, we use path integrals. A path integral is a sum over all paths between two points. It is defined mathematically on pg. 35 of Quantum Mechanics and Path Integrals by Richard Feynman.

$$K(b,a) \; = \; \int_a^b e^{(i/\hbar) \int [b,a]} Dx(t)$$

K is the kernel while a and b are the two points. X is the position between a and b. This definition makes use of the integral within the exponent to find the sum of all paths between a and b, rather than just the area under one path.

On page 62 of his book, Feynman gives us an equation for a moving particle in a potential field. The equation relates the potential energy of the field to the path the article takes, as seen in the equation below. To find the potential one must use the final equation to add all the positions of the path with respect to x and y.

$$V(x)=V(\overline{x}+y)$$

$$V(x)=V(\overline{x})+yV'(\overline{x})+\frac{y^2}{2}V''(\overline{X})+\frac{y^3}{6}V'''(\overline{x})+...$$

It is easier to model the motion of a point in two dimensional space as two moving particles x and X. Feynman shows in detail how this is done mathematically on page 68 of his text. The resulting kernel from this path integral is shown below.

$$K(b,a)= \int_a^b \int_a^b exp\left\{ \frac{i}{\hbar} \int_{t_a}^{t_b} \frac{m}{2}x^2\, dt + \frac{i}{\hbar} \int_{t_a}^{t_b} \frac{M}{2}X^2 dt - \frac{i}{\hbar} \int_{t_a}^{t_b}V(x,X,t)dt \right\}Dx(t)DX(t)$$

Also on page 68, Feynman gives us the result of this integral over the paths X(t). The resulting equation is shown below.

$$K(b,a) \;=\; \int_a^b exp\left\{ \frac{i}{\hbar} \int_{t_a}^{t_b} \frac{m}{2}x^2 dt \right\} T[x(t)]Dx(t)$$

Feynman interprets these results for us saying that "Integrating over all paths available to the X particle produces a functional T." (68, Feynman). The value of a functional depends on a complete function. Feynman further states that, "Thus, the amplitude K, like all the others, is a sum over the amplitudes of all possible alternatives. Each of these amplitudes is a product of two lesser amplitudes." T is this lesser amplitude and is different at points a and b.

As far as this relates to the discussion of telepathy, this integral shows us how the path taken by a particle could influence thought. It provides a basis for analyzing thought particles. Each thought produces a kernel that interacts with the quantum fields produced by the mind in order to transmit information. The mind is the sum of these fields and thoughts are the particles. Telepathy occurs when the particle takes a path from a to b, and we use path integrals to calculate that result.

Telepathy and String Theory

Given the course of modern theoretical physics, it would seem inappropriate to venture a theory that says nothing of string theory. Fortunately, our hypothesis of massless thought particles traveling throughout the universe, perhaps purposefully directed by the thinker, works well within string theory. Massless particles can be treated as strings of energy.

The key is that strings resonate, like guitar strings, at different harmonics. Leonard Susskind writes in his book, The Cosmic Landscape, that, "In principle an ideal infinitely thin string could oscillate in an infinite numbers of harmonics at higher and higher frequency, although in practice, friction and other contaminating influences damp thee vibrations almost before they get started." (225, Susskind) These thought particles can be modeled as infinitely thin strings. The vibrations of these strings could influence the quantum fields of separate minds, hence transmitting information at the speed of light. If higher and higher frequencies are possible, there is no limit to the information that could be communicated.

Susskind also writes that, "All of the possible vibrations, all infinitely many modes of oscillation simultaneously vibrate in a mad symphony of pure noise." This is why I hypothesize that people cannot use telepathy because of the sheer amount of noise dominating the thought scape. It would take a purposeful mind directing its own thought to dampen the noise and reduce it to specific harmonic oscillations and from there perceive another's thinkers thought patterns. Such a talent, would surely be limited to a small number of people. However, it does seem like a skill that can be cultivated by any thinker.

Conclusion:

Let us take a moment to summarize what we have learned thus far. There is extensive experimental evidence that indicates that some people are capable of communicating without visual or auditory cues. It seems quite possible that all people may possess some level of telepathic ability. The very nature of thought itself lends itself to telepathic communication. Our study of telepathy indicates that thoughts themselves are particles while our minds, our sense of self that does the thinking, are quantum fields generated by the neurons sending electrical signals in our brains. The mathematics of quantum mechanics provide ample room for the basics of telepathy, ie dream sharing. Path Integrals provide us with the ability to model and analyze the interaction of these thought particles between minds over great distances. Thus, there is a precedent for telepathy already within physics. String theory adds a layer of beauty that simplifies the concepts and hence makes them all the more acceptable.

There is much room for further research. It would seem that a study needs to be done of energy exchanges through telepathic means in order to find more evidence for the massless thought particles. From there we need to learn how these particles are created.

Overall, I feel confident when I say that telepathy is real and warrants further research and development. Those who think they may be telepathic should not immediately believe that they are insane. It seems quite likely that they are picking up on additional symbols not available to other people. They should learn to direct their thoughts to dampen the noise. Hopefully, that will have the effect of clarifying the stream of information and lead to improved telepathy. Thank you for reading this study.

Works Cited

Feynman, Richard Phillips, et al. Quantum Mechanics and Path Integrals. Dover Publications, 2014.

Susskind, Leonard, and Art Friedman. Quantum Mechanics: the Theoretical Minimum. Penguin Books, 2015.

Susskind, Leonard. "The Cosmic Landscape: String Theory and the Illusion of Intelligent Design." Barnes & Noble, 9 Oct. 1167,

www.ingramcontent.com/pod-product-compliance
Lightning Source LLC
Chambersburg PA
CBHW021854170526
45157CB00006B/2436